39th Biennial Meeting of the American Society of Sugar Beet Technologists 2017

Greenville, South Carolina, USA
27 February – 2 March 2017

ISBN: 978-1-5108-6130-5

Printed from e-media with permission by:

Curran Associates, Inc.
57 Morehouse Lane
Red Hook, NY 12571

Some format issues inherent in the e-media version may also appear in this print version.

Printed by Curran Associates, Inc. (2018)

American Society of Sugar Beet Technologists
800 Grant Street, Suite 300
Denver, Colorado 80203-2944

Phone: 303-832-4460
Fax: 303-832-4468

www.bsdf-assbt.org

Additional copies of this publication are available from:

Curran Associates, Inc.
57 Morehouse Lane
Red Hook, NY 12571 USA
Phone: 845-758-0400
Fax: 845-758-2634
Email: curran@proceedings.com
Web: www.proceedings.com

TABLE OF CONTENTS

CHANDA, ASHOK K.* and JASON R. BRANTNER, Department of Plant Pathology, University of Minnesota, Northwest Research and Outreach Center, Crookston, MN 56716. **Integrated management of Rhizoctonia on sugar beet: resistant varieties, at-planting and postemergence fungicides**

ABSTRACT

Rhizoctonia damping-off and crown and root rot (RCRR) caused by *Rhizoctonia solani* AG 2-2 have been the most common root diseases on sugar beet in Minnesota and North Dakota for several years. Disease can occur throughout the growing season and reduces plant stand, root yield, and quality. Warm and wet soil conditions favor infection. Disease management options include rotating with non-host crops (cereals), planting partially resistant varieties, planting early when soil temperatures are cool, improving soil drainage, and applying fungicides as seed treatments, in-furrow (IF), or after emergence.

In 2015, a field trial was established to evaluate an integrated management strategy consisting of a resistant, moderately resistant and a susceptible variety with seed treatments (penthiopyrad and sedaxane) or in-furrow azoxystrobin alone and in combination with a postemergence azoxystrobin band application for 1) control of early-season damping-off and RCRR and 2) effect on yield and quality of sugar beet. Prior to planting, soil was infested with *R. solani* AG 2-2-infested whole barley (35 kg ha^{-1}). The trial was sown in six-row plots (22-inch row spacing, 30-ft rows) on May 08 at 4.5-inch seed spacing utilizing randomized complete block design. By 5 weeks after planting, plant stands were highest for the resistant variety, followed by the moderately resistant, and lowest for the susceptible variety. In all varieties, seed treated with penthiopyrad or sedaxane, or in-furrow azoxystrobin resulted in higher emergence and stand establishment over 9 weeks than the untreated control. There were significant (P = 0.05) variety by at-planting treatment interactions on yield and significant (P = 0.05) variety by at-planting by postemergence treatment interactions on root rot rating and recoverable sugar A^{-1} (RSA). Resistant and moderately resistant varieties had lower root rot ratings and higher number of harvested roots and RSA compared to the susceptible variety. Plots treated with azoxystrobin in-furrow had lower root rot ratings and higher number of harvested roots, yield, and recoverable sucrose than untreated controls, while plots with penthiopyrad or sedaxane seed treatments were not significantly different from untreated controls. There were no significant differences between untreated control and postemergence azoxystrobin application on any harvest parameters because of dry growing conditions during later part of the growing season in 2015.

In 2016, a field trial was established to evaluate an integrated management strategy consisting of a resistant and moderately susceptible variety with seed treatments (penthiopyrad, sedaxane, and fluxapyroxad) in combination with a postemergence azoxystrobin band application at the 4- or 8-leaf stage for 1) control of early-season damping-off and RCRR and 2) effect on yield and quality of sugar beet. Prior to planting, soil was infested with *R. solani* AG 2-2-infested whole barley (35 kg ha^{-1}). The trial was sown in six-row plots (22-inch row spacing, 30-ft rows) on May 03 at 4.5-

inch seed spacing. Plots were set up in a split-split plot design. Main plots were varieties, the first split was seed treatments, and the last split was postemergence azoxystrobin timings. Each variety by seed treatment combination was planted in triplicate, so that at the 4- or 8-leaf stage, one plot of each variety by seed treatment combination received a postemergence 7-inch band application of azoxystrobin (14.3 fl oz product A^{-1}) while one was left as a stand-alone treatment. There were no significant ($P = 0.05$) two way or three way interactions for stand or harvest data. Resistant and moderately susceptible varieties had similar stands until 3 weeks after planting (WAP), and from 4 to 7 weeks resistant variety had higher stands compared to moderately susceptible variety and by 8 weeks both varieties had similar stands. At-planting (seed) treatments and untreated control had similar stands until 4 WAP and all the three seed treatments had higher stands compared to untreated control from 5 to 8 WAP. Moderately susceptible variety had higher yield and also higher sugar loss to molasses (SLM) compared to the resistant variety. Root rot rating, yield, and other harvest parameters were not significantly different for the seed treatments and untreated control. Root rot severity and percent incidence (percent of roots with a disease rating of > 2.0) was significantly higher in the treatments without postemergence azoxystrobin application, intermediate in 4-leaf azoxystrobin, and lowest in the 8-leaf azoxystrobin application. Yield and recoverable sugar A^{-1} (RSA) was higher for 8-leaf azoxystrobin application compared to no azoxystrobin because of favorable conditions for development of disease later in the season in 2016.

Overall, seed treatments protected early-season stand, but did not result in increased yield, while postemergence azoxystrobin increased yield when conditions later in the season were favorable for disease, but not when conditions were dry. Varieties with at least a moderate level of resistance outperformed the susceptible variety.

HANSON*[1], LINDA E., QIANWEI JIANG[2], HANGHANG WU[2] NOAH ROSENZWEIG, PAULA SAMOHANO, and J.MITCH MCGRATH[1]. [1]USDA-ARS and [2]Michigan State University, East Lansing, MI 48842.

Alternaria leaf spot of sugar beet: factors associated with risk

Starting in 2010, increased incidence and severity of Alternaria leaf spot has been observed in Michigan and similar increased instances are reported in other growing regions. In the past, Alternaria leaf spot in sugar beet was considered a minor foliar disease issue in the United States and management of this disease usually has not been required. If severe, Alternaria leaf spot can cause yield loss due to defoliation as well as to increased sodium and other impurities that interfere with sugar extraction. The aim of the current work was to examine disease management implications in Michigan. Testing included screening Alternaria leaf spot susceptibility in sugar beet germplasm as well as testing the response of *Alternaria* spp. (all in the *A. alternata* species complex) from beet to foliar fungicides. Over 70% of the recent isolates (last two years) were resistant to quinone outside inhibitor (QoI) fungicides, with effective concentration for 50% reduction in growth (EC50) values greater than 60 ppm. Over 90% of the isolates showed tolerance to organotin, with EC50 values between 5 and 10 ppm. Over 50% of the isolates showed resistance to demethylase inhibitor fungicides, with EC50 values greater than 100ppm for some compounds. Resistance to benzimidazole fungicides also was identified in isolates of *Alternaria* from sugar beet. Screening for response to copper fungicides and other fungicides is ongoing. Sugar beet germplasm with differential reaction to *Alternaria* has been identified from field screens and are being tested in the greenhouse with inoculation of lants in a mist chamber, similar to what has been used with Cercospora leaf spot. Infection developed in the mist chamber, but no differences were seen in response of two germplasm that varied in amount of leaf spot observed in the field. Differential responses were observed on detached leaves from the outermost whorl of leaves from various USDA germplasm when inoculated and maintained in a high humidity container. Different age leaves are being tested for their response to Alternaria strains to develop improved inoculation methods.

ANALYSES OF RHIZOCTONIA SCREENING NURSERY RESULTS OVER 15 SELECTED YEARS FROM 1980 TO 2015.

Lee Panella*, Ann L. Fenwick, Travis Vagher, and Mark S. West
USDA, Agricultural Research Service, 1701 Center Avenue, Fort Collins, CO 80526.

Introduction:

The USDA-ARS has had a research program at Fort Collins focused on breeding for resistance to rhizoctonia crown and root rot (Rcrr) since the late 1950s (Pierson and Gaskill, 1961). By 1980, current resistant and susceptible checks were in use. Individual roots from each plot were lifted and rated on a disease index (DI) scale of 0 (healthy, no disease) to 7 (dead, or missing) (Panella, 1998) (Figure 1). Their scores were averaged to provide a plot mean, used for analyses.

Because each year and each experiment is a different environment (temperature, rainfall, position in the field, etc.), one of the largest difficulties faced is comparing DI scores of sugar beet nursery entries across experiments and years. We have examined a number of questions. Do DI scores from susceptible and resistant checks, when compared to DI mean scores from trials, allow us to understand the potential restrictions of DI values with increasing disease pressure? Along those lines we asked how consistent our checks performed over years, experiments, and seed lots. We were very interested in finding good ways to compare results over years and among experiments within years? We also were interested in looking at commercial hybrids over the past 20 years to see if their average resistance had increased or decreased over time. One of the most important purposes of our screening nursery is to identify Plant Introductions with resistance to Rcrr. Therefore we examined if plot variance among individual roots could indicate accessions that may be segregating for resistance to Rcrr, and how we best could compare our results over years, when most accessions are screened in only one year. An important piece of answering and of these questions is to be able to compare the disease severity from experiment to experiment and over environments. We developed a metric that uses the disease ratings to measure the disease pressure (DP)of the environment under which the ratings were taken.

Materials and Methods:

Full data (individual root DIs in plots) from 119 experiments ranging over 15 years between 1980 and 2015 were labeled by type (commercial hybrid, ARS release, Plant Introduction [genebank accession], or seed company experimental line) and source (cooperative, ARS research program, or seed company) (Table 1). Additionally, there were five years, in which we ran all USDA-ARS rhizoctonia resistant releases from the Fort Collins sugar beet research program with the same susceptible check (2000 and 2005 – below, as well as 1999, 2001, and 2002). Resistant and susceptible checks were labeled by name and by seed lot.

Results and Discussion:

In this presentation we focused on a metric to measure the severity of the disease in the nursery and how it impacts the evaluation of disease ratings. We used the commonly occurring highly resistant (FC709-2) and the rhizoctonia susceptible control lines to create a Disease Severity Rating. It can be used to describe an entire year's worth of experiments at one site, an individual experiment or one replicate of an experiment. We used a DI of 3 as the break point because DI scores of 0 to 3 are considered harvestable (Panella, 1998), and looking at a number of different isolates of Rhizoctonia solani of varying aggressiveness, a DI score of 3 was the average break point where the DI was significantly different from the non-inoculated control. The equation used is:

$$DP= (([(PH_R+PH_S)/2 - 0.5]/0.5)*50) + 50$$

Where:

PH_R = the proportion of the roots of the highly resistant control (FC709-2) scored with a DI of 3.0 or more.

PH_S = the proportion of the roots of the susceptible control (FC901/C817) scored with a DI of 3.0 or more.

Multiplying by 50 and adding 50 simply converts the scale from a -1 to 1 scale to a 0 to 100 scale.

Therefore:

DP = 50 represents an ideal environment, i.e., all susceptible plants are scored with a DI of 3.0 or more, and all the highly resistant plants are scored with a DI of less than 3.0.

Table. 1. Full data (individual root DIs in plots) from 119 experiments ranging over 15 years between 1980 and 2015 were used.

	Year	Experiments	Experimental Design (5 replicates)
	1980	8	Randomized Complete Blocked
	1985	9	Randomized Complete Blocked
	1990	10	Randomized Complete Blocked
	2000	8	Randomized Complete Blocked
	2005	10	Randomized Complete Blocked
	2006	9	Randomized Complete Blocked
	2007	7	Randomized Complete Blocked
	2008	8	Randomized Complete Blocked
	2009	8	Randomized Complete Blocked
	2010	9	Randomized Complete Blocked
	2011	9	Randomized Complete Blocked
	2012	10	Randomized Complete Blocked
	2013	9	Randomized Complete Blocked
	2014	2	Randomized Complete Blocked
	2015	3	Completely Randomized
Total	**15**	**119**	

We applied our metric described above by observing the frequency distributions of disease scores relative to the DP metric (Figure 2). Frequency distributions were constructed based on disease scores assigned to plants from those replicates falling into each DP category (low, medium or high). Medium disease pressure was arbitrarily set as the range between 45 and 55. DI scores for susceptible plants tend to be skewed left with a preponderance of scores on the high side when DP is in the medium or high side but tend to be somewhat uniformly distributed among all scores when DP is low. Likewise, DI scores for resistant plants tend to be skewed to the right when DP is medium or low but more uniformly distributed when disease pressure is high. We propose that a DP metric is needed to better understand the separation between resistant and susceptible plants relative to level of disease pressure.

Figure 1. Illustration of the different disease categories used to rate individual sugar beet roots

This page intentionally left blank.

Figure 2. Disease pressure changes the distribution of the resistant and susceptible checks. Frequency tables are a good way to see this. (Medium disease pressure is between 45 and 55.)

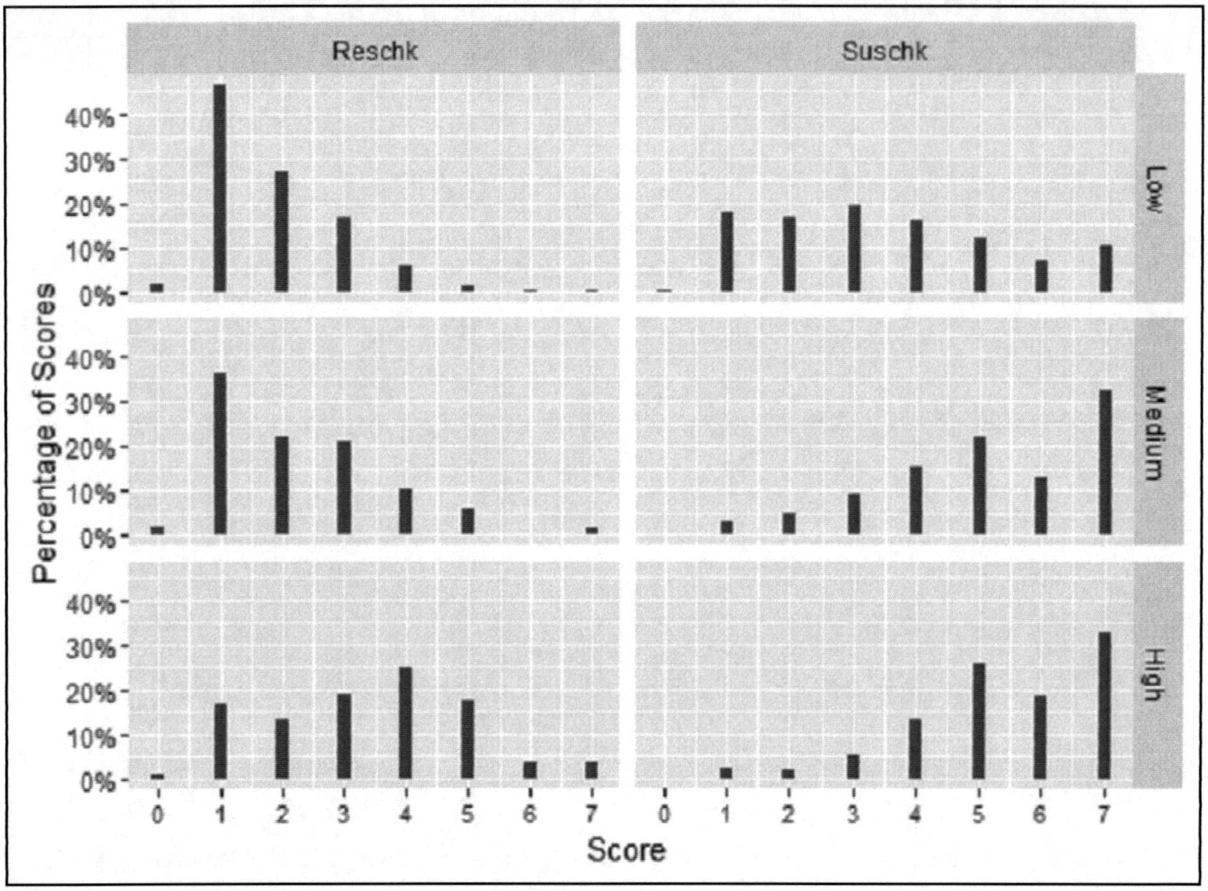

References

Panella, L. 1998. Screening and utilizing *Beta* genetic resources with resistance to *Rhizoctonia* root rot and *Cercospora* leaf spot in a sugar beet breeding program. p. 62-72. *In* L.Frese, L. Panella, H.M. Srivastava, and W. Lange (ed.) Int. *Beta* Genetic Resources Network. A report on the 4th Int. *Beta* Genetic Resources Workshop and World *Beta* Network Conf. held at the Aegean Agric. Res. Inst., Izmir, Turkey, 28 February - 3 March, 1996. International Crop Network Series No. 12. International Plant Genetic Resources Institute, Rome.

Pierson, V. G., and J. O. Gaskill. 1961. Artificial exposure of sugar beets to *Rhizoctonia solani*. J. Am. Soc. Sugar Beet Technol. 11:574-590.

REGITNIG, PETER J.*and BRYAN R. AVISON, Lantic Inc., 5405 – 64th Street, Taber, Alberta, T1G 2C4. **Assessing variety improvements in Alberta.**

ABSTRACT

Historical seed production for Alberta involved an internal company breeding program that continued until 1977. The last variety released from this program was an open-pollinated variety named CS43 that was used until 1980. Several bags of this seed were kept in storage and in 1993 a fresh increase of seed of CS43 was produced in order to compare performance with the best commercial varieties at that time. A small quantity of this 1993 seed was again kept in storage. Germination was 74% when CS43 seed was tested in 2015. A small plot study was conducted in 2015 and 2016 to measure improvements in variety performance by comparing this 1970's variety to a current Roundup Ready variety. CS43 was overplanted and thinned to ensure sufficient stand was achieved in these trials.

Roundup Ready sugar beet varieties were first grown in Alberta in 2009 and exceptional commercial yield and quality have been observed in the last few years compared to previous years when conventional varieties were grown. To make a Roundup Ready comparison with more recent conventional seed, some stored seed of the conventional variety Beta 1385 was used. Beta 1385 was approved for use in Alberta from 2003 to 2009 and seed in storage had a 91% lab germination when tested in 2015. There was also interest in comparing a recent Roundup Ready variety with conventional herbicide to the same variety with Roundup herbicide in order to assess the genetic gains with and without the impact of herbicide setback.

Two trials, each with 8 replications were conducted in 2015 and 2016 with a 1970's variety (CS43), a commercial variety from 2003-2009 (Beta 1385) and a current commercial Roundup Ready variety (HM 9221RR). These trials were planted to a final stand with a modified John Deere 71 Flexi-planter. For the variety CS43, each planter bar used for seeding was filled with 2 times the number of seeds used for the other varieties. In 2015, the emerged stand for CS43 was thinned to the equivalent stand of other treatments in the trial, while lower emergence for CS43 in 2016 did not require any further thinning in that trial.

Broadcast herbicide applications were conducted on all treatments to eliminate the need for inter-row cultivation. The conventional herbicide program in both tests consisted of preemergence Etho, postemergence Betamix β and hand labour. The Roundup herbicide program consisted of postemergence applications of Roundup WeatherMax.

There was a highly significant average gain of 52% in extractable sugar per acre (ESA) when comparing a current Roundup Ready variety (HM 9221RR) sprayed with Roundup WeatherMax herbicide to the 1970's variety (CS43) sprayed with conventional herbicide. When HM 9221RR and CS43 were both sprayed with conventional herbicide the average gain in ESA was 40% for the Roundup Ready variety. Trials conducted in 1994 and 1995 showed a 22% gain in ESA when current commercial varieties were compared to CS43, suggesting similar rates of varietal improvement before and after the mid 1990's when herbicide setback was not taken into account. Substantial plant growth setback was observed for varieties sprayed with conventional herbicides in the 2015/2016 trials and this setback resulted in highly significant ESA reduction. When HM 9221RR was sprayed with Roundup WeatherMax, ESA was 8% higher than the same variety sprayed with conventional herbicides.

There was a highly significant ESA increase of 6% when HM 9221RR with conventional herbicide was compared to Beta 1385. This is in contrast to an even greater 15% increase when

HM 9221RR with Roundup was compared to Beta 1385. These trials suggest that eliminating herbicide setback has had at least as much impact as genetic improvement on increasing ESA since the introduction of Roundup Ready sugar beets.

Considerable improvements have been made in extractable sugar per ton (EST) when comparing current varieties to CS43. There was no difference in EST for HM 9221RR with Roundup or conventional herbicide and these treatments averaged 18% higher than CS43. EST improvements of 6% were observed for HM 9221RR with conventional herbicide compared to Beta 1385. The percent sugar was lower and the molasses loss was higher for CS43 than for other varieties and these differences were highly significant. There was also a highly significant reduction in percent sugar and a highly significant increase in molasses loss for Beta 1385 compared to HM 9221RR.

The increase in root yield when HM 9221RR with Roundup herbicide was compared to CS43 was 7.1 tons/ac or 27%. When HM 9221RR with Roundup applied was compared to the same variety with conventional herbicide there was a 2.3 ton/ac (8%) advantage in root yield. The plant growth setback with conventional herbicide mostly affected root yield and not quality in these trials.

CS43 exhibited significantly lower plant vigor than a current Roundup Ready variety sprayed with Roundup WeatherMax. Herbicide injury was a significant factor in reduced vigor ratings for CS43. A highly significant vigor reduction for HM 9221RR with conventional herbicide compared to HM 9221RR with Roundup WeatherMax herbicide reflected the impact of these two weed control systems on early season plant development. It was also noted that the current Roundup Ready variety appeared visually less susceptible to conventional herbicide injury than the older varieties that were tested.

The tare value for CS43 was 1.1 to 1.6 percentage points higher than for Beta 1385 and HM 9221RR in these trials and these differences were highly significant. Although root morphology was not quantified, it was noted that newer varieties had nicer shaped roots than CS43.

Alberta commercial statistics for the 5-year period from 1975 to 1979 showed root yields averaging 16.4 tons/ac with a sugar content of 16.7% The same measurements for the 5-year period from 2010 to 2014 reported root yields averaging 27.4 tons/ac with a sugar content of 19.0%. Extractable sugar was not measured in the 1970s; however, these differences equate to a 90% increase in gross sugar per acre when root yield and percent sugar are combined. When commercial field production numbers are considered, agronomic improvements must be accounted for along with genetic improvements. Commercial production gains of 90% in gross sugar per acre between 1975-79 and 2010-14 compared to a maximum 43% gross sugar per acre increase in these small plot trials would suggest that a significant percentage of on-farm improvement is related to both agronomic and genetic improvements.

RHIZOCTONIA RESISTANCE CONFERRED BY A SUGAR BEET POLYGALACTURONASE-INHIBITING PROTEIN GENE IN GENETICALLY MODIFIED PLANTS

Haiyan Li[1], Linda E. Hanson[2] and Ann C. Smigocki[1]

[1] USDA-ARS, Molecular Plant Pathology, 10300 Baltimore Ave., Beltsville, MD 20705 and

[2] USDA-ARS, Sugarbeet and Bean Research Unit, East Lansing, MI 48824`

This project is in part supported by a Cooperative Agreement No. 58-1245-4-002N between the U.S. Department of Agriculture and the Beet Sugar Development Foundation.

Abstract

Polygalacturonase-inhibiting proteins (PGIPs) are cell wall leucine-rich repeat (LRR) proteins recognized as having a role in plant defense. PGIPs inhibit fungal polygalacturonase (PG) enzymes that break down the polygalacturonate chain in plant cell walls to initiate disease development. The interaction between PGs and plant PGIPs favors the accumulation of oligogalacturonides which elicit a wide range of plant defense responses. We cloned sugar beet *PGIP* (*BvPGIP*) genes from F1016, a root maggot resistant germplasm, and F1010, a susceptible line in order to define their function in disease resistance mechanisms. *BvPGIP1* was shown to be unique in F1016. *BvPGIP2*, cloned from F1016 and F1010, was shown to differ from *BvPGIP1* in eight amino acids. *BvPGIP1*and *BvPGIP2* were each reconstructed for constitutive expression and introduced separately into *Nicotiana benthamiana.* Independently derived transgenic plants were identified by genomic PCR, Southern blot and RT-PCR analyses. Plants exhibiting high levels of expression were inoculated with *Rhizoctonia solani* AG2-2 isolate R1 or AG4 isolate Rzc27, isolated from sugar beet. Two weeks post-inoculation with fungal-colonized barley grains, stem rot and seedling death were significantly reduced on transgenic *BvPGIP1* and *BvPGIP2* plants as compared to the controls. With AG2-2, seedling death was reduced by 70% - 82% on *BvPGIP1* and *BvPGIP2*, and with AG4, by 50% on *BvPGIP1* plants. No enhanced resistance was seen with AG4 on *BvPGIP2* plants. These initial results suggest that sugar beet *BvPGIP* genes play an important role in plant defense and could provide an approach for controlling fungal diseases in sugar beet and other crops.

Introduction

Plant cell walls constitute the first line of defense against microbes. Polygalacturonase-inhibiting proteins (PGIPs) are plant cell wall proteins that inhibit pathogen and pest polygalacturonases (PGs). Like most of the resistance proteins, PGIPs are leucine-rich repeat (LRR) proteins of the extracytoplasmic type (LxxLxLxxNxLT/SGxIPxxLxxLxx) (Jones and Jones, 1997). They typically contain 10 imperfect LRRs of 24 residues each. In sugar beet, an additional LRR domain, bringing the number to 11, was first reported in the FC607 germplasm (Li et al., 2015). Nine LRRs were reported in *Oryza savita* L. OsPGIP1 and *Medicago truncatula* MtPGIP2 (Janni et al., 2006; Song and Nam, 2005). The amino acids of PGIP that determine specificity and affinity for PGs occur at positions internal to the conserved xxLxLxx motif, which is predicted to form a ß-sheet/ß-turn structure (D'Ovidio et al., 2004; Leckie et al., 1999). PGIP tend to be highly specific and selective for their targets, and a variation of even a single amino acid in the ß-strand/ß-turn motif can confer a new PG recognition capacity (Leckie et al., 1999).

Many plants possess more than one PGIP with differential abilities for inhibiting specific PGs. For example, the four *Phaseolus vulgaris PGIP*s (*PvPGIPs*) have varied specificities against fungal and insect PGs, with *PvPGIP2* encoding the most effective and wide spectrum inhibitor so far characterized (D'Ovidio et al., 2004). Four soybean *GmPGIPs* are differentially expressed after mechanical wounding or infection by *Sclerotinia sclerotiorum* (Kalunke et al., 2014). Recently, three sugar beet *PGIP* genes, *BvPGIP1*, *BvPGIP2* and *BvPGIP3*, were isolated from a sugar beet root maggot (SBRM)-resistant germplasm, F1016. *BvPGIP1* was determined to be unique to F1016. Only *BvPGIP2* and *BvPGIP3* were found in an SBRM-susceptible line, F1010 (Li et al., 2016). *BvPGIPs* were shown to be up-regulated by wounding. In addition, beet line F1016 and F1010 protein extracts inhibited several fungal PGs suggesting that they may play an active role in defense responses (Li et al., 2016).

To determine the role of BvPGIPs in plant defense, *BvPGIP1* and *BvPGIP2* were reconstructed for over-expression in a model plant *Nicotiana benthamiana*. We report on *Rhizoctonia solani* AG2-2 and AG4 resistance conferred by highly expressed sugar beet *BvPGIP1* and *BvPGIP2* genes in tobacco plants, with a differential response for the two genes.

Materials and Methods

Analysis of BvPGIP1 and BvPGIP2 LRR domains

BvPGIP1 and *BvPGIP2* genes were cloned as previously reported (Li et al., 2016). LRR domains in the BvPGIPs were predicted by comparing their sequences with PvPGIP2 for which secondary crystallographic structure is available (Di Matteo et al., 2003).

Plant transformation with *BvPGIP1* and *BvPGIP2*

BvPGIP1 and *BvPGIP2* genes were cloned in Gateway vectors and subcloned into a plant transformation vector (Invitrogen, Carlsbad, CA; http://gateway.psb.ugent.be/). *BvPGIP* genes were fused to the constitutive cauliflower mosaic virus (CaMV35S) promoter for over-expression in transgenic *N. benthamiana* plants.

Total RNA was isolated from plants for RT-PCR analysis (Qiagen, USA; Clontech Laboratories Inc., CA, USA). As previously described, *BvPGIP1* and *BvPGIP2* gene transcripts were amplified using gene specific primers (Li et al., 2016). The housekeeping gene *actin* was used as a control for RT-PCR analysis.

Plant infection with *R. solani*

R. solani AG2-2 isolate R1, highly virulent on sugar beet (Nagendran et al., 2009), and AG4 isolate Rzc27, moderately virulent on beet seedlings, were used for inoculating *BvPGIP1* and *BvPGIP2* transgenic plants. Fungal isolates were grown on potato dextrose agar (PDA) at room temperature. Sterilized barley grains (autoclaved at 120 °C for 20 min, 3 times) were placed in a single layer on a 3 day-old fungal culture on PDA plates and incubated at room temperature for an additional 3-5 days or until all grains were fully colonized. A PDA plate with sterilized barley grains was used as a control to eliminate any other possible source of fungal contamination. Six- to eight-leaf stage tobacco seedlings were inoculated by placing eight fungus-infected barley grains on the soil surface 2-cm away from the seedling stem. Control plants were mock-inoculated with sterilized barley grains in the same manner. Two weeks post-inoculation, the percentage of plants exhibiting seedling death was recorded. Statistical analysis was performed using Student's *t*-test.

Results and Discussion

Comparison of *BvPGIP1* and *BvPGIP2* sequences

Three BvPGIP genes, *BvPGIP1*, *BvPGIP2* and *BvPGIP3*, were isolated from F1016 and F1010, two lines with the most divergent levels of SBRM resistance (Li et al., 2016). *BvPGIP1* was only isolated from the resistant F1016 line (Li et al., 2016). *BvPGIP1* and *BvPGIP2* are highly homologous, differing in only 12 nucleotides at the 3' end (Fig. 1A), producing 8 missense substitutions in their amino acid sequence (Fig. 1B). All 8 amino acids were located in the LRR domain, including 4 internal to the xxLxLxx motif (Fig. 1B), the region considered to be critical for PG recognition (D'Ovidio et al., 2006). Sequence alignments with the *P. vulgaris* PvPGIP2 identified an additional LRR domain (the VII LRR domain) in both BvPGIP1and

BvPGIP2 sugar beet proteins, bringing the total LRR number to 11 (Li et al., 2016). It is unclear what additional function, if any, the longer LRR domains may play in plant defenses as compared to most other PGIPs that characteristically only have 10 LRR domains.

A

```
BvPGIP1    736 ACTCTCCAAACTGTCACTCTTACCGGAACCAATTAAGTGGGCCGGTCCCAACATCGTTG
BvPGIP2        T..............G........A...................................

BvPGIP1    796 GGCCGCGGTAAATTAGCCTTTCTTTATCTTGGTCAAAATAAATTGATAGGAGATGTCTCA
BvPGIP2        ........................................................T...

BvPGIP1    856 TTTTTGTTCCAAAAATCCAACAACAACATCCTCATTTTAGAAATTCAAGACAATCTTTTT
BvPGIP2        ...............A.........T..................................

BvPGIP1    916 AAGTTTGATTTCACAAATGTGGACCTTGCCCCAAGCTTATCAATGTTGAACATGAGCCAT
BvPGIP2        .........................A........................T..A......

BvPGIP1    976 AATATGATATATGGGTCTCTCCCTTTACGATTCGGGCTATTGAACCCGAAT
BvPGIP2        ...............C...........................G....G..
```

B

```
VI   BvPGIP1    224 VVLQLDFNSLSGQIPTSLGLLPTL
     BvPGIP2        ......................S.

VII  BvPGIP1    248 QTVTLTGNQLSGPVPTSLGRGKL
     BvPGIP2        ...A..R................

VIII BvPGIP1    271 AFLYLGQNKLIGDVSFLFQKSNNNI
     BvPGIP2        ....................T....

IX   BvPGIP1    296 LILEIQDNLFKFDFTNVDLAPSL
     BvPGIP2        .......................

X    BvPGIP1    319 SMLNMSHNMIYGSLPLRFGLLNP
     BvPGIP2        ....I..........P.....S.

XI   BvPGIP1    342 NSVDVSYNQLCGPIPNGRR
     BvPGIP2        D..................
     Consensus      xxLxLxx
```

Fig. 1. The difference between *BvPGIP1* and *BvPGIP2* nucleotide and protein sequences. A. The replacement of 12 nucleotides between *BvPGIP1* and *BvPGIP2*, in which 8 (boxed in grey) result in missense substitutions. B. The LRR domain showing the 8 amino acids (in bold) corresponding to the missense substitutions between BvPGIP1 and BvPGIP2. Dots indicate identical sequences. The xxLxLxx motif regions are underlined.

BvPGIP1 and *BvPGIP2* gene expression in *N. benthamiana*

Both of the target genes, *BvPGIP1* and *BvPGIP2* were successfully reconstructed and transformed into N. benthamiana. Four independently derived transformants, three carrying the *BvPGIP1* and one the *BvPGIP2* gene, were regenerated. They were morphologically indistinguishable from normal, untransformed plants (data not shown). A minimum of ten plants originating from each independently-derived T0 transformant were verified by PCR and RT-PCR analysis. In subsequent T2 generations, abundant gene expression was detected in *BvPGIP1* (Fig. 2A) and *BvPGIP2* (Fig. 2B) plants at comparable levels. No mRNA message homologous to *BvPGIP1* or *BvPGIP2* was detected in normal, untransformed plants and the empty vector control (Fig. 2A, 2B).

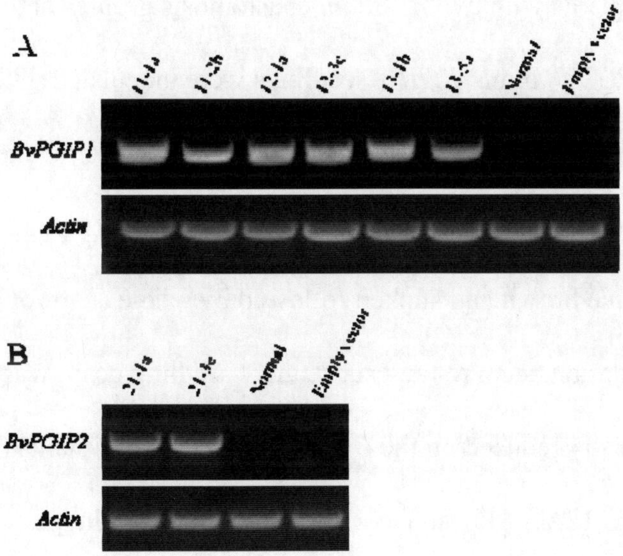

Fig. 2. The expression of *BvPGIPs* in *N. benthamiana* plants. RT-PCR analysis of *BvPGIP1* (A) and *BvPGIP2* (B) in transgenic plants, a normal (untransformed plant) and an empty vector control plant. Independently-derived transgenic plant number 11, 12, 13 and 21 were analyzed twice from two different plants (a and b). The *actin* gene was used as an internal control for transcription levels.

***Rhizoctonia* resistance in *BvPGIP* plants**

The effectiveness of recombinant PGIPs to control fungal growth has been reported in many transgenic plant species. Recombinant PGIPs effectively controlled diseases caused by *Botrytis cinerea*, *Fusarium graminearum*, *Alternaria solani*, *R. solani* and *Bipolaris sorokiniana* (Borras-Hidalgo et al., 2012; Desiderio et al., 1997; Ferrari et al., 2012; Manfredini et al., 2005; Powell et al., 2000). Transgenic tobacco expressing grapevine *VvPGIP1* (Joubert et al., 2006) or expressing bean *PvPGIP2* (Manfredini et al., 2005) showed significantly reduced *B. cinerea* susceptibility. *PvPGIP2* was also shown to significantly increase resistance to *R. solani* AG3 in tobacco (Borras-Hidalgo et al., 2012).

To test whether expression of the sugar beet *BvPGIPs* increases fungal resistance, we inoculated transgenic *N. benthamiana* plants with two *R. solani* anastomosis groups that infect sugar beet and many other plants. *BvPGIP1* and *BvPGIP2* seedlings were inoculated with isolates AG2-2 R1 or AG4 Rzc27. Two to three days post-inoculation, small water-soaked disease lesions were generally observed on stems at the soil level. About one week post-inoculation, these lesions quickly became brown and sunken followed by whole stem rot and seedling death. Two weeks post-inoculation, stem rot and subsequent seedling death was severe on control plants but was limited or greatly reduced on the *BvPGIP1* or *BvPGIP2* plants (Fig. 3). *BvPGIP1* seedling death was reduced to 12% - 41% and to 13% in *BvPGIP2* seedlings inoculated with AG2-2, reducing seedling death by 70% - 82% as compared to the controls (Table 1). Inoculations with AG4 showed reduced *BvPGIP1*seedling death at 25% - 36% which was a reduction of about 50% as compared to the controls (Table 1). No enhanced resistance was seen with AG4 on *BvPGIP2* plants.

Fig. 3. Symptom comparison between the *BvPGIP1* (A) and normal, untransformed control (B) plant two weeks post-inoculation with *R. solani* AG2-2.

Table 1. Percent seedling death (%) of *BvPGIP1* and *BvPGIP2* seedlings inoculated with *Rhizoctonia solani* AG2-2 or AG4.

Transgenes	Transformants	Seedling death (%)*	
		R1	Rzc27
BvPGIP1	11	40.5±13.5[a]	29.2±6.7[a]
	12	14.3±3.5[b]	25.0±1.7[a]
	13	12.1±2.1[b]	35.7±8.4[a]
BvPGIP2	21	13.3±3.9[b]	50.0±7.1[b]
Controls	Normal	73.3±11.8[c]	53.3±8.9[b]
	Empty vector	80.0±10.2[c]	62.9±3.9[b]

*Values represent mean percentage of seedling death ± standard deviation (SD) at two weeks post-inoculation. Experiments were repeated 3 times in replicates of 5 plants from each independently-derived transformant. Means followed by different superscript within columns are significantly different (P<0.05) by Student's *t*-test analysis.

Conclusion

Recombinant expression of sugar beet *BvPGIP1* and *BvPGIP2* genes in tobacco significantly enhanced plant resistance to two isolates of *Rhizoctonia,* representing different anastomosis groups known to cause root rot and damping-off in sugar beet and other plants. Our results indicate that *BvPGIP1* contributed to *R. solani* AG2-2 and AG4 resistance, whereas, *BvPGIP2* conferred enhanced resistance to only the AG2-2 isolate. This could be due to the observed difference in the eight amino acids between these two proteins (Fig. 1). It is well established that a single amino acid substitution can cause changes in PG-PGIP specificity (Leckie et al., 1999; D'Ovidio et al., 2004; Raiola et al., 2008). Our data demonstrates that sugar beet BvPGIPs that have a longer LRR domain function in plant defense, as reported for PGIPs with shorter LRR regions, and that they could provide an approach for fungal disease control in sugar beet and other crops.

References

Borras-Hidalgo O, Caprari C, Hernandez-Estevez I, De Lorenzo G, and Cervone F (2012) A gene for plant protection: expression of a bean polygalacturonase inhibitor in tobacco confers a strong resistance against *Rhizoctonia solani* and two oomycetes. Front Plant Sci 3: 268

Desiderio A, Aracri B, Leckie F, Mattei B, Salvi G, Tigelaar H (1997) Polygalacturonase-inhibiting proteins (PGIPs) with different specificities are expressed in *Phaseolus vulgaris*. Mol Plant-Microbe Interact 10: 852-860

Di Matteo A, Federici L, Mattei B, Salvi G, Johnson KA, Savino C et al. (2003). The crystal structure of polygalacturonase-inhibiting protein (PGIP), a leucine-rich repeat protein involved in plant defense. Proceedings of the National Academy of Sciences of the

United States of America 100: 10124-10128

D'Ovidio R, Raiola A, Capodicasa C, Devoto A, Pontiggia D, Roberti S, Galletti R, Conti E, O'Sullivan D, De Lorenzo G (2004) Characterization of the complex locus of bean encoding polygalacturonase-inhibiting proteins reveals subfunctionalization for defense against fungi and insects. Plant Physiol 135:2424–2435

D'Ovidio R, Roberti S, Di Giovanni M, Capodicasa C, Melaragni M, Sella L, Tosi P, Favaron F (2006) The characterization of the soybean polygalacturonase-inhibiting protein (*Pgip*) gene family reveals that a single member is responsible for the activity detected in soybean tissues. Planta 224: 633-645

Ferrari S, Sella L, Janni M, De Lorenzo G, Favaron F, D'Ovidio R (2012) Transgenic expression of polygalacturonase-inhibiting proteins in *Arabidopsis* and wheat increases resistance to the flower pathogen *Fusarium graminearum*. Plant Biol 14: 31-38

Janni M, Di Giovanni M, Roberti S, Capodicasa C, D'Ovidio R (2006) Characterization of expressed Pgip genes in rice and wheat reveals similar extent of sequence variation to dicot PGIPs and identifies an active PGIP lacking an entire LRR repeat. Theoretical and Applied Genetics 113: 1233–1245

Jones DA, Jones JDG (1997) The roles of leucine rich repeats in plant defences. Adv Bot Res 24: 90-167

Joubert DA, Slaughter AR, Kemp G, Becker JVW, Krooshof GH, Bergmann C (2006) The grapevine polygalacturonase-inhibiting protein (*VvPGIP1*) reduces *Botrytis cinerea* susceptibility in transgenic tobacco and differentially inhibits fungal polygalacturonases. Transgenic Res 15: 687-702

Kalunke RM, Cenci A, Volpi C, Sullivan DM, Sella L, Favaron F, Cervone F, De Lorenzo G,

D'Ovidio R (2014) The pgip family in soybean and three other legume species: evidence for a birth-and-death model of evolution. BMC Plant Biol 14: 189

Leckie F, Mattei B, Capodicasa C, Hemmings A, Nuss L, Aracri B, De Lorenzo G, Cervone F (1999) The specificity of polygalacturonase-inhibiting protein (PGIP): a single amino acid substitution in the solvent-exposed beta-strand/beta-turn region of the leucine-rich repeats (LRRs) confers a new recognition capability. EMBO J 18: 2352-2363

Li H, Padmanaban S, Smigocki AC (2015) Characterization of a *Beta vulgaris* polygalacturonase-inhibiting protein: A defense response gene. Am Soc Sugar Beet Tech Proc 28

Li H, Smigocki AC (2016) Wound induced *Beta vulgaris* polygalacturonase-inhibiting protein genes encode a longer leucine-rich repeat domain and inhibit fungal polygalacturonases. Physiological and Molecular Plant Pathology 96: 8-18

Manfredini C, Sicilia F, Ferrari S, Pontiggia D, Salvi G, Caprari C, et al. (2005) Polygalacturonase-inhibiting protein 2 of *Phaseolus vulgaris* inhibits BcPG1, a polygalacturonase of *Botrytis cinerea* important for pathogenicity, and protects transgenic plants from infection. Physiol Mol Plant Pathol 67: 108-115

Nagendran S, Hammerschmidt R, McGrath JM (2009) Identification of sugar beet germplasm EL51 as a source of resistance to post-emergence Rhizoctonia damping-off. Eur J Plant Pathol 123: 461-471

Powell ALT, van Kan J, ten Have A, Visser J, Greve LC, Bennett AB, et al. (2000) Transgenic expression of pear PGIP in tomato limits fungal colonization. Mol Plant-Microbe Interact 13: 942-950

Raiola A, Sella L, Castiglioni C, Balmas V, Favaron F (2008) A single amino acid substitution in highly similar endo-PGs from *Fusarium verticillioides* and related *Fusarium* species affects PGIP inhibition. Fungal Genet Biol 45: 776-789

Smigocki AC, Ivic-Haymes S, Li H, Savic´ J (2013) Pest Protection Conferred by a *Beta vulgaris* Serine Proteinase Inhibitor Gene. PLoS ONE 8(2): e57303

Song KH, Nam YW (2005) Genomic organization and differential expression of two polygalacturonase-inhibiting protein genes from *Medicago truncatula*. J Plant Biol 48: 467-478

SUGARBEET RESPONSE TO SEED POSITION RELATIVE TO FERTILIZER BAND IN A STRIP TILLAGE SYSTEM

W. Bart Stevens*, Jay D. Jabro and William M. Iversen
USDA, Agricultural Research Service, NPARL, 1500 North Central Avenue, Sidney, MT 59270

Introduction:

Strip tillage (ST) is a conservation tillage system that is adaptable to the small-seeded sugarbeet (*Beta vulgaris*, L.) crop (Evans et al., 2010; Overstreet, 2009) as means to reduce fuel consumption and protect the soil. As interest in ST for sugarbeet production increases, growers adopting this technology need information regarding optimum fertilizer management. Most ST equipment currently available is equipped to apply fertilizer in a band below the seed row. Banding N and P can improve fertilizer uptake efficiency because nutrients are close to the developing root system. Placement directly under the seed has been shown to be particularly beneficial for sugarbeet since early root growth is almost entirely downward rather than lateral (Anderson and Peterson, 1978). However, when ST and planting occur in separate operations it can be difficult to consistently plant seed in the optimal position relative to the fertilizer band. If the sugarbeet seed is placed too far to the side of the fertilizer band, early-season nutrient deficiency may result (Stevens et al., 2007). Our objective was to evaluate sugarbeet plant growth and yield when the seed row is offset laterally relative to the fertilizer band.

Materials and Methods:

The study was conducted from 2011 to 2013 on a sandy loam soil in western North Dakota using overhead sprinkler irrigation. Strip tillage was performed in the late October of each year using a semi-parabolic shank type implement that tilled 30-cm wide strips spaced 60 cm apart in barley (*Hordeum vulgare* L.) crop residues. Urea (145 kg N ha^{-1}) and monoammonium phosphate (12 kg N and 56 kg P_2O_5 ha^{-1}) fertilizers were placed in a single band at a 10-cm depth as part of the ST operation. No in-season fertilizer was applied.

Sugarbeet seed was planted in late April or early May to a depth of 3 cm and 0, 7.5, 15, or 22.5 cm to the side of the fertilizer band. A control treatment with no fertilizer applied was also included. Irrigation water was applied based on crop water use so as to maintain soil water content between 50 and 100% of water holding capacity. Weeds were controlled with two to three in-crop glyphosate (N-(phosphonomethyl)glycine, isopropylamine salt) applications.

Whole plant dry matter was determined at approximately 40 days after emergence (DAE) by randomly removing six plants from each plot and washing them to remove soil. Plants were then dried at 60°C for approximately 7 days and weighed to determine plant dry matter (PDM). Harvest root yield was determined by hand-harvesting roots from 3.0 m of row. Fresh weight and sucrose content of harvested roots were determined by the sugarbeet quality laboratory at Sidney Sugars, Inc. and used to determine sucrose yield.

Data were analyzed based on a randomized complete block design with year and seed row position as fixed variables. Block (n=6) and its interactions were treated as random variables. Least-square means for treatment effects on 40-DAE PDM and final harvest yield components were separated using a 95%

($p<0.05$) confidence level. Sucrose yield (kg ha^{-1}) was calculated by multiplying fresh root yield (Mg ha^{-1}) by root sucrose content (kg Mg^{-1}).

Results and Discussion:

In all three years, PDM at 40 DAE was from 0.6 to 3.0 g $plant^{-1}$ when seed was placed 22.5 cm to the side of the fertilizer band (Fig. 1). This was indistinguishable from the control suggesting no significant fertilizer uptake occurred during the first 40 days of plant growth. Excessive precipitation in 2011 leached N from the upper soil profile root zone and limited fertilizer uptake through 40 DAE regardless of the distance between the seed row and fertilizer band. In 2012 and 2013, PDM at 40 DAE was greatest when sugarbeet was seeded directly over the fertilizer band (0-cm) and was only marginally (9 to 15%) lower when offset by 7.5 cm. When the seed row was offset by 15 cm, PDM at 40 DAE was 20 to 55% less (2012 and 2013, respectively) than when the offset distance was 0 cm. These results show the importance of fertilizer placement to early season plant growth and canopy development. Milford et al. (1980) showed that potential yield is strongly correlated to canopy development because greater leaf area intercepts a larger amount of radiation which is then converted to dry matter and sucrose. These researchers also emphasized the importance of early-season expansion of leaf area as a determining factor of leaf potential. Under adequate soil moisture availability and favorable temperatures, N concentration in the leaves is the primary factor determining the rate of early season leaf expansion (Milford et al., 1985). This is clearly illustrated in Fig. 1 as PDM, and presumably leaf area, is much less when the sugarbeet roots are too far away from the fertilizer band to fully access N early in the growing season.

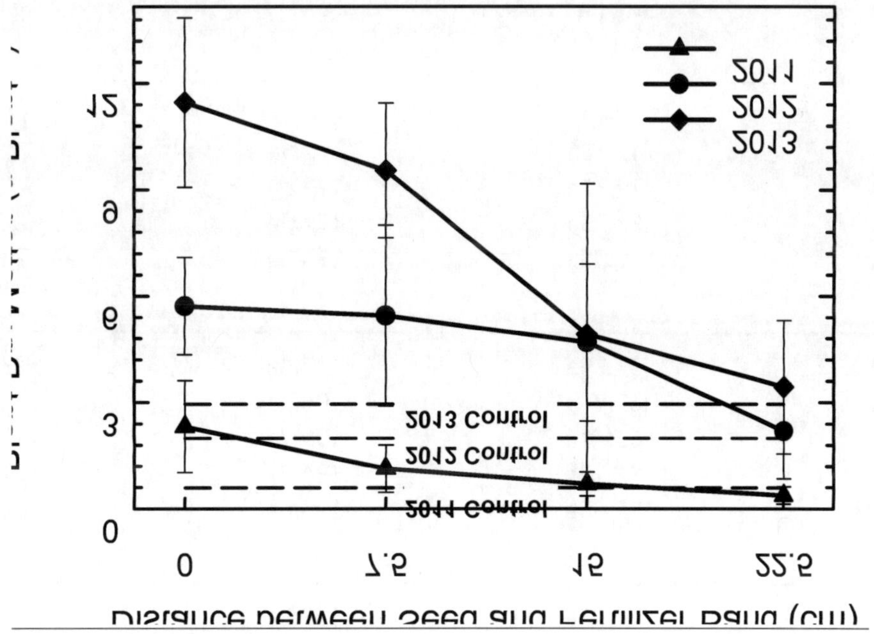

Figure 1. Whole plant dry matter as affected by lateral distance between seed row and fertilizer band in a strip tillage sugarbeet system. Plant dry matter for the control where no fertilizer was applied is indicated by dashed lines. Error bars indicate one standard deviation of the mean.

As was the case with PDM at 40 DAE, the effect of higher than average precipitation in 2011 affected both fresh root yield and sucrose yield dramatically. The highest sucrose yield observed in 2011 was 58.8 and 61.4% of the highest yield observed in 2012 and 2013, respectively (Table 1). In 2012, end-of-season sucrose yield was similar (12262 to 12367 kg ha^{-1}) for the 0-, 7.5- and 15-cm offset distances but was approximately 17% lower for both the 9-cm offset distance and control treatment. In 2013, sucrose yield was similar for the 0- (11841 kg ha^{-1}) and 7.5-cm (10847 kg ha^{-1}) offset distances and decreased by 19, 23 and 43% (compared to the 0-cm offset distance) for the 15- and 22.5-cm offset distances and control treatment, respectively. This increased sucrose yield at the 0- and 7.5-cm offset distances is likely the result of more rapid canopy development due to greater N uptake during the early growing season as compared to seed placement >7.5 cm away from the fertilizer band (Fig. 1). This improved canopy development undoubtedly led to more sunlight being harvested and greater sucrose production where early season N availability was greatest. These results agree with those of Stevens et al. (2007) who showed that under conventional tillage, sucrose yield was greater when preplant N applications were placed 8 cm from the seed row than when placed 18 cm away.

Table 1. Effect of lateral distance between seed and fertilizer band on recoverable sucrose yield components in sugarbeet grown using strip tillage from 2011 to 2013. Control was tilled but no fertilizer was applied.

Distance	Root Yield‡	Root Sucrose Content‡	Sucrose Yield
cm	Mg ha^{-1}	kg Mg^{-1}	kg ha^{-1}
	------ 2011 ------		
0.0	40.03a§	172.1a	6869a
7.5	42.18a	172.5a	7276a
15.0	39.31a	172.1a	6760a
22.5	37.39a	174.5a	6530a
Control	40.95a	175.2a	7169a
	------ 2012 ------		
0.0	62.79a	197.4a	12376a
7.5	61.87a	198.7a	12288a
15.0	63.41a	195.7a	12262a
22.5	52.19b	194.6a	10166b
Control	54.05b	193.6a	10424b
	------ 2013 ------		
0.0	72.78a	162.8a	11841a
7.5	65.56a	165.3a	10847a
15.0	56.72b	166.9a	9442b
22.5	55.06b	164.5a	9054b
Control	39.85c	163.4a	6532c

‡Root yield and sucrose content are for fresh weight roots at harvest moisture content.
§For a given year, means in a column followed by the same letter are not significantly different (*p*<0.05).

Conclusions:

Distance between seed and banded fertilizer has a pronounced influence on sugarbeet yield due in part to the relationship between early season N availability and canopy development. When the separation between fertilizer and seed is excessive, seedlings may suffer from early-season N deficiency. Consequently, dry matter accumulation, including leaf expansion, is slowed resulting in a reduced ability to optimally harvest sunlight and convert solar radiation to sucrose. We conclude that for ST sugarbeet production, the lateral distance between sugarbeet seed and the fertilizer band should be 7.5-cm or less for optimum early plant growth and harvest yield.

References:

Anderson, F.N., and G.A. Peterson. 1978. Optimum starter fertilizer placement for sugarbeet seedlings as determined by uptake of radioactive 32P isotope. J. Am. Soc. Sugarbeet Techol. 20:19-24.

Evans, R.G., W.B. Stevens, and W.M. Iversen. 2010. Development of strip tillage on sprinkler irrigated sugarbeet. Appl. Eng.- Agric.26: 59-69.

Milford, G.F.J., P. Biscoe, K.W. Jaggard, R.K. Scott, and A.P. Draycott. 1980. Physiological potential for increasing yields of sugar beet. In *Opportunities for Increasing Crop Yields,* pp. 71-83. Eds R.G. Hurd, P.V. Biscoe, and C. Dennis. London: Pitman.

Milford, G.F.J., T.O. Pocock, J. Riley, and A.B. Messem. 1985. An analysis of leaf growth in sugarbeet. IV. The expansion of the leaf canopy in relation to temperature and nitrogen. Ann. Appl. Biol. 107:335-347.

Overstreet, L.F. 2009. Strip tillage for sugarbeet production. Intern. Sugar J. 111:292-304.

Stevens, W.B., A.D. Blaylock, J.M. Krall, B.G. Hopkins, and J.W. Ellsworth. 2007. Sugarbeet yield and nitrogen use efficiency with preplant broadcast, banded, or point-injected nitrogen application. Agron. J. 99:1252-1259.

Stevens, W.B., R.G. Evans, J.D. Jabro, and W.M. Iversen. 2011. Sugarbeet productivity as influenced by fertilizer band depth and nitrogen rate in strip tillage. J. Sugarbeet Res. 48:137-154.

WINTERMANTEL, WILLIAM M. [1]*, KIMBERLY M. WEBB[2], AND JESSICA L.S. RUPP,[3]
[1]USDA-ARS, 1636 East Alisal St., Salinas, CA 93905, [2]USDA-ARS, 1701 Centre Ave., Ft. Collins, CO 80526, and [3]Montana State University, Bozeman, MT 59717. **Advantages of using 'omics' technologies and bioinformatics for analyzing the impact of pathogens on sugarbeet.**

ABSTRACT

Throughout the history of American sugarbeet production, research has proceeded hand-in-hand with the emergence of new diseases, and sugarbeet scientists have used the technologies available to improve disease management and crop yield in the face of emerging disease pressures. Many traditional methods will always remain important for disease management, including pathogen-host range studies, pathogen isolation, inoculation of sugarbeet to confirm causative nature of the pathogen, and identification of environmental factors that contribute to disease. In the late 20th Century the advent of molecular biology revolutionized plant pathology. The development of antiserum-based detection methods gained prominence for detection of viruses, and staining and plating facilitated detection of bacterial and fungal pathogens. In the 1980s and 1990s, DNA and RNA based detection methods, and the ability to rapidly identify pathogen variants due to unique characteristics of their DNA were rapidly developed for the detection and quantification of all types of pathogens. Like these methods which are now common within the research community, the advent of the genomics era is again revolutionizing sugarbeet research.

The development of methods for analysis of the complete sequence of the genome (total DNA), transcriptome (RNA produced from DNA), proteome (all expressed proteins), and metabolome (total biochemical compounds produced) now allow scientists the ability to essentially begin putting together a systemic analysis of how sugarbeet responds to pathogen infection as well as to other factors. An organism's *genome* is where all of its biological information is stored, and is composed of all the DNA of the organism, including the genes encoded by that DNA. *Genomics* is the study of an organism's genome, including its genes and their structure and function. It can include gene mapping, genome sequencing, and connected interactions within the genome. A *transcriptome* is the entire collection of RNA produced from an organism's genes. The process of making RNA from DNA is called transcription and is an enzymatic process that occurs in each cell. Each gene serves as a template for making a unique RNA. Different amounts of RNA are produced from each gene and this is a highly regulated process. Not all genes are expressed at the same time or in the same way. *Transcriptomics* looks at the timing and the way in which genes are turned on and off (or produce increased or decreased levels of RNA) based on environmental or developmental influences. This can also include changes in response to infection by pathogens, such as viruses, fungi, bacteria, nematodes, or even insects. A *proteome* is the full complement of proteins produced by an organism. Just as genomic DNA is used as a template for making RNA, the RNA is used as a template for making proteins by a cellular process known as translation. Essentially the RNA serves as a code for the synthesis of protein. This process is called "translation." *Proteomics* is the study of proteins produced by an organism, and like transcriptomics, looks at the timing of protein production, amounts of protein produced, and even protein regulation (making more or less of specific proteins). A *metabolome* is the complete set of metabolic molecules produced by an organism. Metabolites are usually small molecules that are involved in growth, development, reproduction, or other secondary processes. Metabolites usually have important functions and their expression reflects events affecting the organism, much as one sees with protein or RNA. Primary and secondary metabolites often act as compounds that regulate the functional pathways of proteins and/or RNA. *Metabolomics* is the study of these molecules, including the timing and level of their expression, usually in response to specific stimuli.

While the technology associated with "omics" is complex, the end result of studies using these methods can help decipher how sugarbeet responds to pathogen attack. Studies by our laboratories have addressed changes in sugarbeet using proteome and metabolome analysis to understand the effects of Rhizomania and Rhizoctonia crown and root rot on sugarbeet. Others have used transcriptomic and genomic analysis. Each of these methods is connected to the others and all utilize information encoded by the genome. The real advantage of such methods is that they can be used to obtain both knowledge of gene expression by healthy sugarbeet as well as when sugarbeet is impacted by a pathogen or influenced by external environmental factors at the molecular level. By evaluating a pathogen-host or pathogen-vector system using multiple strategies, such as transcriptomics and proteomics or proteomics and metabolomics, for example, it is possible to determine changes in an organism in response to pathogen attack and gain a high level of confidence as to which pathways are altered during pathogen infection. This information can be used with genetic modification approaches to prevent pathogen infection, or or alternatively could be used to enhance production by improving performance of select functions of the sugarbeet.

Two primary methods are now available to utilize knowledge gained through the genome-based methods described above, RNA interference (RNAi) and CRISPR/CAS9. RNAi is a biological process in which small RNA molecules are used to induce changes in RNA levels in an organism. RNAi works by triggering a natural process that eliminates specific RNAs from an organism. This can alter gene expression (RNA or protein levels) and change the manner in which an organism such as sugarbeet responds of stimuli, including pathogens (i.e. virus control) or in some cases herbicides (Roundup Ready™). CRISPR-Cas9 is a newer and more targeted molecular biology method for editing specific sequences within a genome without altering other parts of the genome. A cell's DNA can be cut in a specific location allowing genes or parts of genes to be removed and/or replaced with other sequences. Like RNAi, CRISPR has application for control of pathogens or to introduce new and improved traits in sugarbeet and other organisms. As noted above, these methods could also be applied for trait enhancement as well as pathogen control. Molecular markers have been in use for many years, but with the availability of complete genome sequences, these too can be used even more efficiently to identify and move select traits within breeding populations.

Genomic, transcriptomic, proteomic, and metabolomic information can be generated within a relatively short period of time, and can facilitate targeted strategies, leading to improved pathogen control. By integrating omics technologies into traditional sugarbeet pathology, genetics, and breeding, we can advance our ability to protect sugarbeet from disease and environmental stresses, leading to improved and more reliable production.

WINTERMANTEL, W.M.[1]*, L.L. HLADKY[1], A.A. CORTEZ[1], L.-F. CHEN[2], and R.L. GILBERTSON[2], [1]USDA-ARS, 1636 East Alisal Street, Salinas, CA 93905 and [2]University of California, 1 Shields Ave., Davis, CA 95616. **Differential accumulation and transmission of traditional and emergent *Beet curly top virus* strains from of the western United States.**

ABSTRACT

Curly top disease, caused by various strains of *Beet curly top virus* (BCTV), causes significant economic losses for sugarbeet throughout the western United States. Recent surveys in California and the Pacific Northwest identified a shift in the predominant strains of BCTV impacting agriculture, with newly emergent strains supplanting traditional strains of the virus in both regions over the past few years. This has resulted in some changes in severity and affected host plants that may reflect the prevalence of the new strains over traditionally common strains. Prior to the recent shift in virus population structure, BCTV strains CFH and Worland (formerly known as *Beet severe curly top virus* and *Beet mild curly top virus*, respectively) were widely prevalent in sugarbeet production throughout the West, with limited detection of the California/Logan strain. In California, three strains, CO95, LH71, and a *Pepper curly top virus* isolate, BV3, have emerged and displaced CFH and Worland. In Idaho, a slightly different variant of CO95 has emerged as well and has been the most prevalent isolate in that region over the past two years.

To identify factors leading to emergence of new strains, competition studies were conducted in which sugarbeet and other host plants were inoculated with newly emergent and traditional BCTV strains as single or co-infections using agroinoculation of cloned viral DNA. Some infections were established using leafhopper transmission due to difficulty in obtaining viable infection with clones of specific strains using the agroinoculation method. After four weeks, test plants were evaluated for virus accumulation using quantitative PCR (qPCR) with TaqMan probes specific to known prevalent strains of BCTV, including CFH, Worland, LH71, CO95, and BV3, the latter of which was previously found associated with curly top symptoms on sugarbeet in the San Joaquin Valley of California. Following confirmation of virus titer, plants with single and mixed infections of BCTV strains were used for leafhopper transmission to new test plants. This involved 48 h acquisition and transmission periods, with 15 leafhoppers per plant, after which plants were allowed to develop infections. After four weeks, qPCR was used to monitor accumulation of each strain in test plants to determine which strains were transmitted most effectively.

Results demonstrated preferential accumulation of some strains in a host plant-specific manner. Previous studies had shown that the Worland strain accumulates well in bean and shepherd's purse plants, but is less competitive in sugarbeet compared with the CFH strain. Current studies demonstrated that BV3 accumulates less effectively in both sugarbeet and tomato than does CFH, the strain that until a few years ago was often the most prevalent and severe isolate found in sugarbeet production regions. When the two viruses were present in mixed infections results were similar to those from single infections, and this was also reflected in lower transmission rates of BV3 from mixed infections than for CFH. The LH71 strain also failed to accumulate well in sugarbeet during single infections, but accumulated quite well in tomato and shepherd's purse plants. In mixed infections, however, the LH71 strain did not accumulate as effectively in tomato or sugarbeet, but again was competitive with CFH in shepherd's purse, and was transmitted by leafhoppers to a greater degree than was CFH. Competitition studies involving CO95, arguably the most important of the newly emergent strains in both California and Idaho, was hampered by inefficient transmission by the clone used to establish standardized infections. CO95 is genetically similar to the Worland strain, and like Worland, is less efficient at establishing infections using agroinoculation. Therefore studies on competitiveness and host preference of CO95 are continuing but at a slower rate due to the need to study this virus entirely with leafhopper transmission. Finally, continuing studies are examining accumulation of individual isolates from weed hosts common in Idaho and California, from which leafhoppers carrying BCTV migrate to fields of sugarbeet and other crops each spring. Once completed, this information should clarify the role of key host plants in the epidemiology of the newly emergent BCTV variants in these important agricultural regions, and should contribute to decisions on virus management and weed control.

In summary, results demonstrate that certain BCTV strains are better adapted to some host plants than others, and that host plants appear to be exerting a substantial influence on emergence and dominance of the new BCTV strains. Secondarily, studies also found that specific BCTV strains (Worland strain and those relatively closely related to the Worland strain) have difficulty in establishing infections from cloned DNA using agroinfection, a common method of virus inoculation in laboratories, and this clearly influences virus accumulation during mixed infections putting these strains at a selective disadvantage in establishing infections. Results have implications for virus propagation for curly top nurseries, and begin to clarify factors influencing emergence of new BCTV strains.

AUTHOR INDEX